FM SATELLITE COMMUNICATIONS FOR BEGINNERS

SHOOT FOR THE SKY... ON A BUDGET

BRIAN SCHELL

Written and designed by:
Brian Schell
brian@brianschell.com

Version Date: September 5, 2018

ISBN: 9781720105855

Printed in the United States of America

CONTENTS

INTRODUCTION

Sputnik launched in 1957. Earth's first artificial satellite was a milestone for humanity, and it orbited the Earth for three entire weeks before its batteries died, and it burned up as it fell back to Earth. Modern satellites have improved a bit since then in longevity, power, and price. Rather than requiring a "space race" and the resources of nations, satellites and other payloads are launched on a regular basis, many of which are communications related.

Still, for most people, the workings of satellites are a little mysterious, and they're often the source of unfounded paranoia. What do they do up there? Are they watching us? Can they fall on my house? Even today, these kinds of questions are still too common, as the masses direct suspicion at the mysterious government agencies and powerful corporations who control most of the satellites.

Many of us can remember, "back in the day," that many news broadcasts had the subtitle "Live via satellite" below the foreign correspondents. This was somewhat after "Now in COLOR" stopped being a thing. There was a time when satellite TV was a novelty that required a six- or eight-foot dish and cost so much that only the TV stations could afford to have them. Now, small "Dish TV" antennas are literally everywhere, and in many cases are cheaper to use than running a cable.

Nearly everyone has used GPS in some fashion or watched satellite TV. Yet satellite technology still feels like a new and complicated thing, literally out of reach for most people. And using a satellite directly? Other than NASA experts and "rocket scientists," who does that?

Amateur radio enthusiasts do it, that's who!

OSCAR 1, the first amateur satellite, was launched way back in 1961, just four years after Sputnik. Now, there are numerous radio-equipped satellites in orbit that cater specifically to hams. They're free to use, and the equipment required is easy to obtain and set up. The historically difficult part of the operation is knowing where the satellites are (where to point the antenna), and that's easily solved today with a simple web search or a smartphone app.

There are varying levels of difficulty in working with satellites. Some are easy contacts, others are harder to hit. Sometimes you can reach a destination thousands of miles away, sometimes you can talk to an astronaut on the ISS. There are as many goals and projects in the satellite segment of the hobby as there are with any other "Ham subculture."

Let's see how this all works.

WHY DO THIS?

The purpose here is much the same as any other facet of amateur radio: to get more from technology, to communicate with other amateurs, and because it's a fun challenge. There are sometimes situations where there is no HF radio available, or conditions are unsuitable for HF use, and long-distance calls are desired. With a satellite setup, you can use a simple and inexpensive handheld radio to talk to very distant stations.

The basic idea of what we're going to do is this: Use a radio and a directional antenna to receive and transmit signals to and from orbital satellites. There's a ton of science and history behind all this, but we're going to focus specifically on the what-you-need-to-know details of how to do this. Learning the history and science of satellites is for your own research.

These satellites, in turn, do different things. Some simply transmit beacons. Others contain transponders and digi-peaters that will relay your communications to other amateurs on the ground. The most common activity is using a satellite as transponder: Your signal goes into space, hits the satellite, then gets retransmitted down to someone else on the ground, just like a repeater in space. The most popular ham satellites are "cubesats" or small 10-centimeter cubes that are essentially repeaters in space.

That said, it's not all about just talking to other people on the ground. You can also use the same technology to talk to astronauts aboard the International Space Station, and even download SSTV images from the space station as well. No matter where your radio interests lie, there are a large number of options available in the satellite portion of the hobby that may appeal to you.

EQUIPMENT

Talking to or through a satellite sounds like it ought to be a really expensive thing to do, but it's really a lot simpler than you might think at first. Only a few things are required, and many of those can be built or improvised.

Radio

It might seem obvious that you're going to need a radio, but there are some specific things to be aware of.

Most satellites use two different frequencies. It's very common for a satellite to transmit on the 2m band and receive on the 70cm band (or vice versa), so you will definitely need either a dual-band radio or two single-band radios. Satellite communications are very low-power, so handheld (HT) radios are perfectly adequate to the job. Moreover, they are a lot more portable than a mobile radio. Still, if a mobile is what you have, you can use that well enough too. Once you have mastered the FM satellites, you may want to move up the difficulty scale to SSB transmissions through linear transponders, and that in turn necessitates more complex equipment.

It's not strictly necessary, but it's nice if your radio is "full duplex," meaning it is capable of transmitting and receiving at the same time. Most handheld radios are _not_ full duplex, but that's not a deal breaker. The difference is that with full duplex, you can hear yourself being transmitted through the satellite, so it's a great way to know everything is actually working; there are a lot of variables, and just being confident that you're actually hitting the satellite makes your job a lot easier.

If you aren't planning on talking (just listening), you can use any scanner that allows you to tune into the receive frequency. You'll just set the scanner frequency to whatever the satellite transmits on, plug in the antenna, and aim in the proper direction at the correct time. You don't even need to be a licensed amateur radio operator to just listen. Doing it that way is only half the fun, but it works!

Antenna

It's completely possible to talk to a satellite using the rubber duck antenna on your HT radio; people do it all the time. On the other hand, it's not really something you can depend on, and getting through is more of a matter of luck and having the weather conditions be just right than any real skill. You'll do much better with an antenna designed for this.

Most satellite folks swear by small Yagi antennas, and the two most popular varieties are the Arrow II and Elk antennas. People have been arguing back and forth over which model is better for decades, and there's no clear winner. I have an Arrow II (see the cover of the book— that's me and my Arrow), and that works great for me, but others swear on their Elk. They're roughly the same price ($150-ish), so although they aren't cheap, this is really the only purchase most hams need to make to get into this area of the hobby.

One interesting complication with this is that since satellites work on two bands at once as mentioned above, you transmit on 2m and receive on 70cm. And this requires a dual-band radio or two single-band radios. In order to use a dual-band radio, the antenna needs a *diplexer* to allow one radio to control both bands with a single antenna jack.

I'm not much of a do-it-yourselfer, but it's not especially difficult to make your own directional antenna. Just Google "ham radio satellite antenna plans" and there are dozens of links to show you how. I recently watched a YouTube video called "$4.00 Ham Radio Satellite Antenna" where he makes one from a stick and some wire coat hangers. I've seen more than one article online about making one from an old tape measure. There's a link in the "Links and Further Reading" section towards the end of the book. I'm sure it's not going to be as efficient as a purchased antenna, but it looks like a fun project.

Other Items

The radio and antenna are the only absolutely required items to do this, but a few other things will make your life easier and make satellite work a lot more comfortable and fun.

- **A Watch or accurate timer.** To be super-accurate,

use a phone that can check time online. One iPhone app that I've used for years is **Emerald Time**. It uses Network Time Protocol to display the current time of day as defined by the international standard atomic clocks. Its accuracy is limited only by network latency and is usually accurate to less than +/-100 milliseconds. You really don't need anything close to that kind of accuracy for this kind of work, but... why *not* have it?

- **A Flashlight.** You can obviously do satellite work during the day, but it's a lot of fun to go out at night and try to visually spot satellites. Be prepared for darkness. *Note*: You can't actually see any of the ham radio satellites as they are too small, but there are hundreds of others that you *can* spot.

- **A Compass.** You can "get by" with just knowing which way is North, but since most listings will give you very specific compass directions, it's good to know exactly where to face. You can buy an inexpensive "Boy Scout"-style compass easily enough, but don't forget many smartphones have a perfectly good compass app on them.

- **Headphones.** Many satellite signals are very weak, and depending on where you're working, external noises can be quite high. Even in the so-called "quiet countryside," wind and other natural noises can often be a lot louder than you expect.

- **Handheld microphone.** This is more of a convenience than a necessity, but if you're holding the antenna in one hand it's often less of a hassle to use a handheld microphone rather than deal with orienting and finding the transmit button on a handheld radio. If you're using a mobile radio rather than a handheld variety, then this isn't even optional.

- **Recorder.** Most people hold the antenna in one hand and the microphone in the other. How are you going to write down callsigns, times, and other notes? You can't write them down, so a simple voice recorder is a really helpful device. Again, a cheap device is fine, and again, a smartphone voice-recorder app could be all you need.

- **Tripod.** For improved aiming accuracy and to ease arm strain, a tripod with some form of clamp to hold your antenna is going to make your life easier. This may not be the best way to get started (you can actually learn a lot by flailing your antenna about in the beginning), but once you've made a few contacts, this is something to try. I've even seen some hams make brackets to hold both the antenna and the radio. They have gotten to the point of almost being hands-free.

- **Computer, Tablet, or Phone.** In the previous section, I mentioned using smartphone apps for a compass, voice recorder, timer, and so forth, but the most important use is for looking up where the satellite is going to be at a specific time. This can all be done ahead of time at home, and it's probably best to look ahead and find out what time satellite X is going to be in a good position.

Again, if you're in a hurry, or have to get by on a small budget, most of the accessories above can be skipped or built/bought at a later date if you need. If you already have a smartphone, many of the tools can be replaced with the proper apps.

THE PROCESS

Research What You Can Do

This method works on any device that can access the web. We'll talk about platform-specific apps and software later, but for now, let's use what's freely available on websites.

Using Web Tools

With any satellite tracking tool, first you'll need to know either your precise latitude and longitude, or more easily, your grid square. If you don't know your grid square, you can find it here:

http://www.levinecentral.com/ham/grid_square.php

Once you have that, go to the AMSAT page at:

http://www.amsat.org/track/

CN 87 RH

Amsat Start

In my case, I know my grid square is en83da, so I can plug that in and it will figure out my latitude and longitude. If you know your elevation above sea level, put that in too, but it's not crucial.

Last, look the options in the drop-down box for "Show predictions for" which defaults to showing "ISS." You can choose from roughly two hundred different satellites now. Let's say we want to try to hit the AO-85 satellite. Simply choose "AO-85" from the list, check one more time to make sure your location is as accurate as possible, then hit the "Predict" button. It'll come back showing a table that looks something like this:

AO-85
145.980 MHz
DOWNLINK

AMSAT Online Satellite Pass Predictions - AO-85
View the current location of AO-85

Date (UTC)	AOS (UTC)	Duration	AOS Azimuth	Maximum Elevation	Max El Azimuth	LOS Azimuth	LOS (UTC)
11 Jul 18	17:40:28	00:08:13	138	5	112	64	17:48:41
11 Jul 18	19:17:00	00:14:01	201	57	118	38	19:31:01
11 Jul 18	20:58:01	00:13:30	251	23	308	28	21:11:31
11 Jul 18	22:41:50	00:10:25	299	8	339	28	22:52:15
12 Jul 18	00:25:23	00:09:36	329	6	8	48	00:34:59
12 Jul 18	02:06:26	00:13:25	334	16	30	93	02:19:51
12 Jul 18	03:46:43	00:15:38	327	65	75	140	04:02:21
12 Jul 18	05:27:41	00:13:38	312	20	258	190	05:41:19
12 Jul 18	18:02:59	00:11:55	167	17	107	49	18:14:54
12 Jul 18	19:41:45	00:14:17	222	63	300	32	19:56:02

Your results are shown above
Use the form below to request more pass predictions

Show Predictions for: AO-85 ▼ for Next 10 ▼ Passes

Amsat Result

Now it's time for a couple of definitions.

AOS stands for Acquisition of Signal (or Satellite). AOS is the time that a satellite rises above the horizon of an observer.

LOS stands for Loss of Signal (or Satellite). LOS is the time that a satellite passes below the observer's horizon.

UTC is Coordinated Universal Time. It's a standard time zone that can be used anywhere. If you aren't sure what time it is in UTC, check here:

https://www.timeanddate.com/worldclock/timezone/utc

AZIMUTH is the direction you need to face. Start out by

facing north. North is "zero" degrees. Now turn to the right (Eastward) until you have gone the specified number of degrees. If you need 90 degrees azimuth, you'll turn to face due East; 270 degrees would have you facing West. This is where a compass might come in handy.

ELEVATION is the height off the ground that you need to raise your eyes. Zero degrees elevation is looking at the flat horizon. Ninety degrees Elevation is straight up. If, for some reason, you have a negative elevation, that means the target is below the horizon. It may or may not be obvious, but a satellite has to be above the horizon for you to be able to access it.

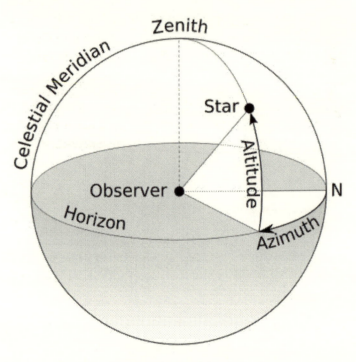

Altitude and Azimuth [source:
http://commons.wikimedia.org/wiki/File:Azimut_altitude.svg]

Now an example using the first line from the Amsat table from a few pages back:

As I'm writing this section, it is July 11th, so if I go outside at 17:40 UTC time, then turn to 138 degrees azimuth at the horizon (the *actual* horizon, not the tree line!), I'll be looking right at AO-85. Over the next 8 minutes and 13 seconds, it will drift to azimuth 64 degrees with the highest point being at 112 degrees azimuth and 5 degrees elevation. If you aren't confident in your ability to judge the elevation angles, pick up a cheap protractor at an office supply store and use that to help visualize the positions.

Be aware, that just because a bird shows up on the chart, that doesn't guarantee you'll have a good chance of hitting it. In the example above, 5 degrees elevation way is too low to be practical in reality, but that's the first line on the chart.

In a real-world case where I wanted to actually HIT the AO-85, I'd probably try to use the information from the *second* line, where the maximum elevation is 57 degrees. At 19:17, I would face 201 degrees azimuth (somewhat south-south-west) and face the horizon. I would be aware that over the next 14 minutes and 1 second that the satellite would drift upwards to a peak at 57 degrees height around azimuth 118, then head back down toward the horizon at 64 azimuth. I'd expect to lose signal completely right around 19:31. Due to the higher altitude and longer time above the horizon, this satellite has a much better chance of being hit successfully than the first example.

If I was patient and wanted the best chance of hitting the bird, the 7th line has the maximum elevation and the longest duration, so that would be the pass I'd look at first.

Of course, depending on the location and any obstacles that might be in the way in a certain direction, there are other factors to take into consideration. Planning is key to success in this.

Sigh. That's a lot of degrees and numbers. It's obviously best to scout out the location ahead of time, get there a little early, and look for landmarks near the start, end, and midpoints of the sat's route.

Side note: For a "just for fun" activity at night sometime, make a list of satellite passes during the time you'll be in a dark place, then see how many you can actually spot visually. There are a lot more satellites that are visible than are available for ham radio use, so you can stay pretty busy identifying all the bright lights that pass overhead. Some people make a hobby just out of visual sightings of satellites; check out the site https://heavens-above.com/main.aspx

But we aren't done yet. You'll also need to know the frequencies (plural) of whatever satellite you want to contact. The most commonly used ham satellites that cover the USA (as of mid-2018) are the SO-50, AO-85, AO-91, and AO-92, but there are some other options for those outside the USA, and several new satellites are scheduled for launch later in 2018. A quick Google search can tell you updated information, but Amsat.org will have the information on the most common satellites.

https://www.amsat.org/fm-satellite-frequency-summary/

And you'll get a long listing that looks something like this:

FM Satellite Frequency Summary

AMSAT Fox-1 Satellites			
	Uplink FM (67 Hz CTCSS)	Downlink FM	Comments
AO-85 (Fox-1A)	435.170 MHz	145.980 MHz	Operational
AO-91 (RadFxSat / Fox-1B)	435.250 MHz	145.960 MHz	Operational
AO-92 (Fox-1D)	435.350 MHz & 1267.359 MHz*	145.880 MHz	Operational
Fox-1Cliff	435.300 MHz & 1267.300 MHz*	145.920 MHz	Launch By End of 2018
* Switchable by command station. Not operational simultaneously.			

Satellite Frequencies and details

So let's look at AO-85 for an example. According to the website, it has an uplink frequency of 435.170 and a downlink of 145.980.

Uplink is the frequency you **transmit** on.

Downlink is the frequency you **listen** on.

The practical problem now becomes how do you transmit on one band and listen on another? There are two common approaches. First, you can use two handheld radios, one for listening, and one for transmitting. There's nothing at all inherently wrong with this approach, but it's more hardware to deal with and just one more thing to mess up. Two radios also means more cabling and more physical "stuff" to arrange and find a place for. The second approach, and the one that most hams use, is to use a single dual-VFO handheld radio with VFO A set to the uplink frequency, and VFO B set to the downlink frequency.

Note also if the satellite you want to use requires a PL tone or a special "wakeup" tone. Some do, some don't.

Compensating for the Doppler Effect

Now the part that gets tricky. You will be standing still, so your position isn't going to change on Earth. The satellite, on the other hand, is moving through space at an incredible speed relative to you. You've heard the change of pitch a fast car makes as it passes while you stand near the road? That's called the "Doppler Effect," and it can have a huge effect on radio frequencies.

The uplink (your transmitting) frequency in the chart for AO-85 is **435.170**. This is the actual frequency the bird uses, but due to its high velocity, the number is going to change as it passes because of the Doppler shift. The best way to deal with this is to initially set the frequency a little **below** the listed frequency to start, and adjust it upwards as it passes over. To save confusion later, many hams will program the frequencies into their radio ahead of time. Of course, the frequencies listed below are only for the AO-85 satellite; each satellite will have a different sequence of frequencies:

> *435.160 (start listening here, a little bit below the*
> *   listed frequency)*
> *435.165*
> *435.170 (the actual frequency listed on the chart)*
> *435.175*
> *435.180 (probably won't get above this, but it*
> *   might get that far)*

So you start transmitting at the lowest frequency on the list, and as the bird passes overhead, you turn to the next frequency on the list, and so forth. Depending on the length of time the satellite is above the horizon, the amount of "doppler drift" could be significant. If the satellite is coming straight toward you, more or less right overhead, then you may need all these frequencies; if it's moving at a low angle, then all five frequencies probably won't be needed. Compensating for Doppler Shift is an acquired skill, and you'll need practice to be able to do it correctly.

After you've done this a bit, and have made contacts on a few different satellites, you may want to program your radio with a "bank" of satellite frequency ranges that you've used. As a rule of thumb, program frequencies ±3.5 kHz for 2m frequencies and ±10 kHz for 70cm frequencies. There are a number of completed example programming setups shown later in the book.

Now here's another tricky bit. You only need to do these adjustments on the 70cm band (that is, the higher frequency band). It doesn't matter which band is the uplink or which is the downlink, but the 70cm band always gets the adjustments. The Doppler shift on the 2m band is little enough that you can disregard it.

In the example above, the **uplink** frequency was 435.170, so we programmed a sequence of uplink frequencies above and below it. On the other hand, if we were programming for the SO-50 satellite, for example, its **downlink** is 436.795, and so that's the number we need to adjust for Doppler Shift. It's easy to remember that **the higher frequency band numbers are always the ones to need adjustments.**

Thought that was complex? Here's one more thing: **Uplink**

**frequencies get adjusted upwards during communica-
tions, while downlink frequencies get adjusted downwards.** If we can use that moving car analogy again, you
know that if you are listening to a fast car drive by, you can
hear the pitch change as it approaches. It's the trickiest part
of the whole project, but we'll go over some very specific
programming examples when we start to discuss the
common satellites later.

Doppler Shift Summary:

- The frequency in the 70cm band needs to be
  adjusted during the satellite pass.
- Uplink frequencies get adjusted upwards
- Downlink frequencies get adjusted downwards.
- Only one 2m frequency is needed. These don't need
  steps, as Doppler shift on 2m is negligible.

Polarity

You have probably seen a photo of a satellite in space, but a
still photo makes a satellite look a lot more stable than it
really is. These little cubesats tumble and roll around a lot,
so "this end up" isn't really a thing in orbit. Those little
antennas are rotating as well, and this sometimes causes
issues with polarity. There's no way to predict or completely
eliminate this problem, but if you have a handheld antenna,
try rotating it and twisting it in your hand until you find the
strongest signal.

Squelch

If you haven't figured it out yet, satellite communication is very hit-or-miss, with a lot of very weak signals and noise. It's best to turn your radio's squelch completely off and listen through the static rather than try to eliminate it. Many satellite transmissions are weak enough that they won't make it through even a low squelch setting.

OTHER SOFTWARE TOOLS

iOS Tools

For the iPhone and iPad, I recommend the HamSat HD, ProSat HD, and ISSHD apps, all created by Craig Vosburgh. All three work in similar ways, to read in your location data and tell you in several different ways what you can access and when. My favorite mode is a 3D globe of the Earth with various satellites shown above it, including different colored circles around each one displaying the radius where it should be accessible; if your map location is within the circle, you might want to give it a try.

All these satellites and circles move and update in real time, so if you move too slowly, you'll be able to see that visually as well. If you click on a specific satellite, you can see the latitude and longitude of the satellite's current location as well as the azimuth and altitude for your own location. The three apps work very similarly to each other, and none are free.

As of this writing, the ISS Edition is $2.99, the HD edition is

$7.99, and the Pro HD edition is $9.99; they are nice-looking but also don't offer anything you can't find on the AMSAT websites. It's just a lot more visual, which my or may not be worth it for you. The ISS edition includes only the International Space Station, while the other two are more for communications satellites, with the main difference being that the PRO version has more satellites (and prob-ably weaker contacts). There are other satellite tracking programs in the App Store, but most of them aren't really oriented toward amateur radio.

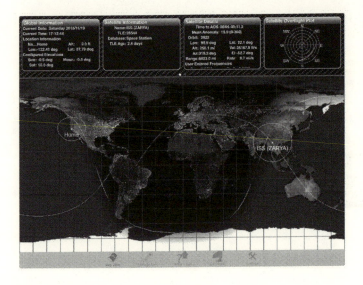

HamSat HD

iPhone Tools

The three apps listed above for iPad are also available for the iPhone. They're universal apps, so if you purchase the app for one device, it'll work on the other, so that's good for planning at home on your big-screen iPad and to carry with you on your iPhone.

Android Tools

Tythatguy has an excellent tracking app called surprisingly "Satellite Tracker" and *RunaR* has a good one for just the ISS called "ISS Detector Satellite Tracker."

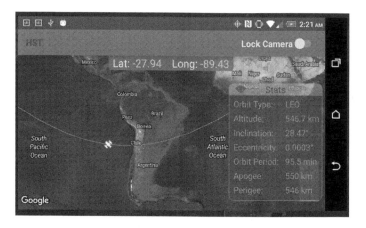

"Satellite Tracker"

"ISS Detector Satellite Tracker"

Windows, Mac, and Linux Tools

My own preference is to use a mobile app like one of the ones already described or the free, web-based apps available on the AMSAT page. Still, there are some very advanced software tools available. Check out the latest releases at the AMSAT page before continuing:

http://www.amsat.org/amsat-new/tools/software.php

Here are some screenshots of the best-looking ones:

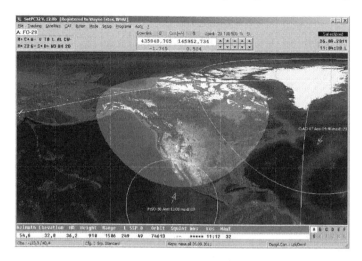

SATPC32 Software for Windows

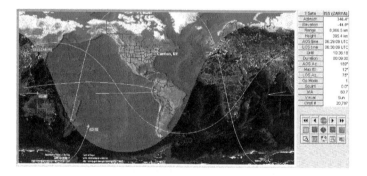

Nova Software for Windows

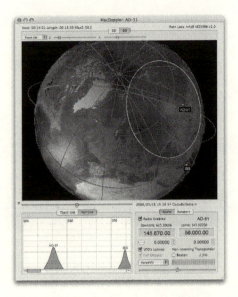

MacDoppler for Mac

USING COMMON FM SATELLITES

There are an uncountable number of satellites in orbit, but only a handful use frequencies that are within the ham bands, and only a few of them are usable with FM. Here are the "usual suspects" for users in North America:

Name	Uplink	Downlink	PL Tone
AO-85	435.170	145.980	67 Hz
LilacSat2	144.350	437.200	None
IO-86	435.880	145.880	88.5 Hz
AO-91	435.250	145.960	67 Hz
AO-92	435.250	145.880	67 Hz
SO-50	145.850	436.795	67 Hz (74.4 wakeup)

Common Satellite Uplinks and Downlinks

The above is quick reference, check out current information before programming your radio:

https://www.amsat.org/two-way-satellites/

Some of these satellites require more in the way of programming considerations. For example, SO-50 has the same 67Hz PL tone as several of the others, but has a "sleep timer" that puts the satellite to sleep after ten minutes. You'll need to "wake it up" by transmitting a 74.4 Hz tone for two seconds before calling CQ. Having TWO PL tones is unusual in that you will need to program two channels into your radio; one for the wake-up call, and one for the transmit frequency. Other weirdness, such as the LilacSat-2 only being available three days a week, and IO-86 only being accessible from equatorial regions, also must be taken into consideration. Whichever bird you want to use, do some research on the AMSAT site to make sure you have all your bases covered.

Some Notable Hamsats

OSCAR 1 (No longer in service)

The first amateur satellite, simply named OSCAR 1, was launched on December 12, 1961, barely four years after the launch of world's first satellite, Sputnik I. The beginning of this project was very humble. The satellite had to be built in a very specific shape and weight, so it could be used in place of one of the weights necessary for balancing the payload in the rocket stage. OSCAR 1 was the first satellite to be ejected as a secondary payload (with Discoverer 36 as the primary) and subsequently enter a separate orbit. The satellite carried no on-board propulsion and the orbit decayed quickly. Despite being in orbit for only 22 days, OSCAR 1 was an immediate success with over 570 amateur radio operators in 28 countries forwarding observations to Project OSCAR.

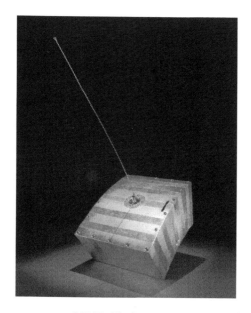

OSCAR-1 The first Hamsat

The following descriptions are paraphrased from Amsat.org:

AO-85 (Operational)

Fox-1A, AO-85 or AMSAT OSCAR 85 is an American amateur radio satellite. It is a 1U Cubesat, was built by the AMSAT-NA and carries a single-channel transponder for FM radio. The satellite has one rod antenna each for the 70 centimeter and 2 meter bands. To enable a satellite launch under NASA's Educational Launch of Nanosatellites (ELaNa) program, the satellite continues to carry a Penn State University student experiment (MEMS gyroscope).

The satellite was launched on 8 October 2015 with an Atlas V rocket together with the main payload, Intruder 11A (also known as NOSS-3 7A, USA 264 and NROL 55) and 12 other Cubesat satellites (SNaP-3 ALICE, SNaP-3 EDDIE, SNaP-3 JIMI, LMRSTSat, SINOD-D 1, SINOD-D 3, AeroCube 5C, OCSD A, ARC 1, BisonSat, PropCube 1 and PropCube 3) from Vandenberg Air Force Base, California, United States. After just a few hours, the transponder was put into operation, initial connections were made between amateur radio stations and telemetry was received.

AO-91 (Operational)

Fox-1B, AO-91 or AMSAT OSCAR 91 is an American amateur radio satellite. It is a 1U Cubesat, was built by the AMSAT-NA and carries a single-channel transponder for FM radio. The satellite has one rod antenna each for the 70 centimeter and 2 meter bands. Fox-1B is the second amateur radio satellite of the Fox series of AMSAT North America.

AO-92 (Operational)

Fox-1D, AO-92 or AMSAT OSCAR 92 is an American amateur radio satellite. Fox-1D is a 1U CubeSat developed and built by AMSAT-NA. Fox-1D carries a single-channel transponder for mode U/V in FM. Fox-1D has an L-band converter (the AMSAT L-band downshifter experiment), which allows the FM transponder to be switched on an uplink in the 23 centimeter band.

SO-50 Satellite

SO-50 (Operational)

SO-50 carries several experiments, including a mode J FM amateur repeater experiment operating on 145.850 MHz uplink and 436.795 MHz downlink. The repeater is available to amateurs worldwide as power permits, using a 67.0 Hertz PL tone on the uplink, for on-demand activation. SO-50 also has a 10 minute timer that must be armed before use. Transmit a 2 second carrier with a PL tone of 74.4 to arm the timer.

The repeater consists of a miniature VHF receiver with sensitivity of -124dBm, having an IF bandwidth of 15 KHz. The receive antenna is a 1/4 wave vertical mounted in the top corner of the spacecraft. The receive audio is filtered and conditioned then gated in the control electronics prior to feeding it to the 250mW UHF transmitter. The downlink antenna is a 1/4 wave mounted in the bottom corner of the spacecraft and canted at 45 degrees inward.

LilacSat-2 (Operational)

The LilacSat-2 FM repeater is currently planned to be active for 24 hours at a time, starting around 2200 UTC on Mondays, Wednesdays, and Fridays. This schedule is flexible, and subject to change without prior notice.

As with other satellites, full-duplex operation is preferred. This allows you to hear your own signals from the satellite, without relying on others to confirm that. Please listen around the 144.350 MHz uplink frequency before working LilacSat-2 passes, as this is a frequency outside the normal 2m amateur satellite sub band at 145.800-146.000 MHz. The 144.350 MHz uplink frequency is a legal uplink frequency per the International Radio Regulations and (in the USA) FCC Part 97, but may conflict with local or regional band plans.

Specific Programming Examples

Back when we first discussed Doppler Shift, I mentioned it was complicated. Let's make it simpler. Here are some specific programming tables incorporating doppler adjustments for four of the most common satellites as well as the ISS. These are actual examples as I have them entered in the CHIRP programming software. If you aren't familiar with CHIRP, check my book on the free software in the "Also By" section. Otherwise, these tables are easily adapted to any other amateur radio programming software:

Memory Range: 0	- 999	Refresh	Special Channels	Show Empty
Loc ▼	Frequency	Name	Tone Mode	Tone
19	0.000000		(None)	
20	435.160000	Acquisition of Signal	Tone	67.0
21	435.165000	Approaching	Tone	67.0
22	435.170000	AO-85 MAIN	Tone	67.0
23	435.175000	Departing	Tone	67.0
24	435.180000	Loss of Signal	Tone	67.0
25	0.000000		(None)	
26	145.980000	AO-85 Downlink	(None)	
27	0.000000		(None)	

AO-85 Programming in CHIRP

As you can see in this first listing, the downlink is at 145.98 mHz (memory channel 26 in this listing), and that's all you need for the 2m band. The "main" uplink frequency is 435.17, and I've programmed two small steps above and below that frequency (memories 20-24). Also, I've included the necessary PL Tone of 67 Hz for all the uplinks to activate the repeater.

In the field, I set my VFO-A to the Downlink frequency, and VFO-B to the first of the uplink frequencies (435.16 in the example). If you're using two single-band handheld radios instead of a dual-band system, then program one radio for the uplink frequencies and the other for the downlink.

If I then point the antenna in the right direction at the right time, I should eventually hear something. In this case, the downlink doesn't change, but the uplink will need to be adjusted upwards as it passes from the point of acquisition of signal to the place where the signal is lost. How much you adjust it depends on how much of an angle the satellite is traveling at in relation to you.

Always wait to make sure you can clearly hear the satellite before you transmit to it. If it seems to be working, but no one seems to hear you speaking, try adjusting to one of the other uplink frequencies.

To program the radio for the AO-91 and AO-92 satellites, the process is pretty similar:

Loc ▼	Frequency	Name	Tone Mode	Tone
0	0.000000		(None)	
1	435.240000	Acquisition of signal	Tone	67.0
2	435.245000	Approaching	Tone	67.0
3	435.250000	AO-91 MAIN	Tone	67.0
4	435.255000	Departing	Tone	67.0
5	435.260000	Loss of Signal	Tone	67.0
6	0.000000		(None)	
7	145.960000	AO-91 Downlink	(None)	

Memory Range: 0 - 999   Refresh   Special Channels   Show Empty

AO-91 Programming in CHIRP

Loc ▼	Frequency	Name	Tone Mode	Tone
9	0.000000		(None)	
10	435.340000	Acquisition of Signal	Tone	67.0
11	435.345000	Approaching	Tone	67.0
12	435.350000	AO-92 MAIN	Tone	67.0
13	435.355000	Departing	Tone	67.0
14	435.360000	Loss of Signal	Tone	67.0
15	0.000000		(None)	
16	145.880000	AO-92 Downlink	(None)	

Memory Range: 0 - 999   Refresh   Special Channels   Show Empty

AO-92 Programming in CHIRP

The SO-50 bird is a little bit of an oddball. The uplink is in the 2m band, and the downlink is 70cm, so everything is a little backwards. In this case, you don't adjust the uplink; instead, you listen for a weakening downlink signal, and when the signal starts to fade, you adjust the downlink frequency to the next step. The SO-50 also needs to have an additional memory programmed because it has a special "wake-up" PL tone that it needs to hear before it activates:

Loc ▼	Frequency	Name	Tone Mode	Tone
29	0.000000		(None)	
30	145.850000	SO-50 Uplink "Wake Up"	Tone	74.4
31	145.850000	SO-50 Uplink	Tone	67.0
32	0.000000		(None)	
33	436.805000	Acquisition of Signal	(None)	
34	436.800000	Approaching	(None)	
35	436.795000	SO-50 MAIN	(None)	
36	436.790000	Departing	(None)	
37	436.785000	Loss of Signal	(None)	

Memory Range: 0 - 999   Refresh   Special Channels   Show Empty

SO-50 Programming in CHIRP

Lastly, here are some frequencies to get you started with contacting the International Space Station. There's no PL tone for anything on the ISS, but depending on which ITU region you are in, the uplink changes:

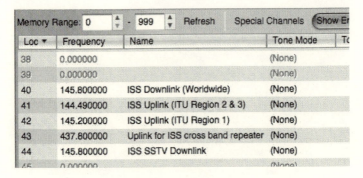

ISS Programming in CHIRP

Where do I get all this information? It's all on the Amsat.org website. It's just a matter of knowing what information you need to prepare before grabbing your antenna.

Satellite Pass Predictions:

https://www.amsat.org/track/index.php

Current Satellite Status:

https://www.amsat.org/status/

Satellite Schedules:

https://www.amsat.org/satellite-schedules/

Specific Details on individual Satellites:

https://www.amsat.org/two-way-satellites/

And

https://www.amsat.org/fm-satellite-frequency-summary/

PREPARATION CHECKLIST

Research

Before you do anything else, look up the information about the satellites you want to try. From Amsat.org, you need:

1. AOS (acquisition of signal) time. Convert this in your head to local time, and double-check this to make sure you don't arrive an hour too soon!
2. Compass direction for where the satellite will rise over the horizon.
3. LOS (loss of signal) time.
4. Compass direction for where the satellite will disappear over the horizon.
5. Maximum altitude of the satellite.
6. Frequency and PL code(s) to program into your radio.

Radio Preparation

1. Make sure squelch is turned off.
2. Program in uplink frequency.
3. Program in a range of doppler-adjusted downlink frequencies (five frequencies usually will do it).
4. Set one band for the uplink, and the other on whichever downlink frequency you are going to start with.
5. Assemble the antenna.
6. Make sure you have any cables or batteries that you may need.
7. Don't forget optional accessories like a compass, recorder, notepad and pencil, flashlight, and anything else that may come up.

On-Site Preparation

1. Look at your compass, and note landmarks where the satellite will appear above the horizon.
2. Also note where it will sink below the horizon.
3. Where will the highest point of the path be.
4. Plug in the antenna to the radio and get your hardware ready.
5. A couple of minutes before the satellite is scheduled to rise, turn on the radio and point the antenna in the right direction, just above the horizon.
6. Wait and listen. Rotate the antenna from time to time; adjusting the polarity sometimes helps with weak signals.

7. Once you hear the bird coming through the downlink, wait for an opening and call CQ, or answer one if someone else is calling.

Logging Your Calls

For a lot of hams, building up the logbook is a big portion of the fun. I wouldn't recommend taking along a computer unless you have a table set up, but somehow you need to keep a record of who you heard, who you talked to, and what times things happened. Assuming you are working alone, the best tool for this job is to simply record everything and sort it out later. If you have someone else along with you to help, this may be less necessary, but signals are often weak enough that it may be hard to catch everything. So even with a helper, a recording may be useful.

Once you get home and can get comfortable, fill up your log with all those juicy contacts. While you are at it, be sure to report to AMSAT what satellite you used and how well it worked through their "status" page:

http://www.amsat.org/status/

TIPS

- If you cannot hear anything through the downlink, do *not* try transmitting to the uplink. Just because you can't hear the bird doesn't mean someone else isn't using it. Transmitting when you can't hear a response just ruins things for everyone.
- Don't overpower the satellite. Five or ten watts is all this should take, and any more than that causes problems - some sats even filter out signals that are too strong.
- Another tip that could help if you have the equipment for it: A device like the SDRplay that can show an entire band visually on a computer screen is extremely useful for fine-tuning your antenna aim. It's probably not necessary to get this precise with tuning, as the birds are moving at high speed and part of the fun is in "the chase" anyway, but it's something to think about if you have an SDR device.
- There's not much time for a QSO, so work as quickly and efficiently as possible. Just call CQ

quickly with a "KD8OTD CQ KD8OTD" to give your call sign twice. Use the regular letters and numbers, don't slow it down with phonetic alphabet. Alternately, to call someone else (maybe arranged ahead of time), give a quick "KA9SCF this is KD8OTD" Once you do make contact with someone, give your name, grid square, and QTH. A signal report is unnecessary-- you're hearing them, so that's all that matters! The bottom line is that lots of people want to make contacts, so hit them quick and let go.

- If you screw up and get something wrong, don't worry about it. Stay relaxed and enjoy the hobby. There's a lot of factors involved with satellite communications, and you may not succeed on the first try (or first few tries). People who start a QSO often drop out for one reason or another.

- Remember, it's as much a challenge for the guy on the other end of the call as it is for you— Things are going to fail.

- It's not a requirement, but remember that a full-duplex radio lets you verify that you are correcting for Doppler Shift properly, that you are hitting the satellite, you have enough power, and that nothing else is wrong. If you can hear your own voice being transmitted from a sat, you know you have everything set correctly. If no one answers, you can be reasonably confident that it's not due to your mistake.

- Don't get beat up or arrested. That sounds like a joke, but it's not. I've heard more than one story of some ham being accused of pointing their antenna near someone's house and having the police called.

It's not hard to explain to someone that you're not "listening" to them, but only if they want to listen. Common sense should make it obvious that those "metal sticks" can't magically "hear" what's going on in your house, but try explaining that to an irate, drunken husband at midnight sometime. You might have better luck explaining to the police, but depending on the situation and timing, maybe not.

- Pick your site carefully. Unless your neighbors all know what you are doing, a small suburban backyard might not be the best place to try any of this. In my case, there's a large disc-golf course, a former full-size golf course, a block from our house, so I just go there and set up. I can point my Arrow antenna any direction I need without worrying, and it's a wide-open space offering plenty of visibility for my own safety. Scout out a place in the daytime before arriving for an after-dark flyover. It's common sense, but enough hams get in trouble over this to make it worth a mention. Even without the risks of getting assaulted or arrested, there's still a matter of tripping or falling over something in the darkness.

- For some specialized digital modes, such as SSTV from the IIS, you generally only receive a broadcast. All you need to do once you've got your equipment set up is to record the audio from the ISS and you can run that through decoding software at home later - it doesn't need to be done in real time!

NEXT STEPS: SSB AND LINEAR TRANSPONDERS

The purpose of this book is to get you up and running with **FM satellites**, but most of the same basic concepts apply to SSB transmissions as well. This section is here just to point out that SSB and FM communications have some practical differences.

Linear transponders are a step or two up in complexity, as most transponder uplinks are in LSB mode and downlink in USB mode, which completely changes the type of radio you'll need. I'm not aware of any current HTs that do SSB, so at the minimum, you'll need a dual-band radio that can do 2m and 70cm in SSB. *Most* mobile UHF/VHF radios can't do SSB either, so the choice of radio models is somewhat small. Also, since this kind of setup often requires a table and power supply, simply going out into a field with a handheld radio is generally not an option. Folks who are into this mode often stay home and use a computer-aimed antenna on a tower. As you can guess, this is a big step up in cost as well.

To make matters even muddier, Doppler shift is far more

complicated with SSB. With FM radios, you can program little steps and adjust the frequency that way, but SSB is so narrow-band that smooth and continuous adjustments are required. Most people who do this kind of work use computer-controlled radios and maybe computer-aimed antennas as well.

There are benefits to using this more complicated system. Multiple users and conversations can take place simultaneously over the range of frequencies covered by these birds, and they also can support CW, APRS, SSTV, PSK31, and many other digital modes.

For a quick look at the satellites you can hit in this manner, and the appropriate modes and frequencies, check out https://www.amsat.org/linear-satellite-frequency-summary/

LINKS AND FURTHER READING

Know your grid square:
http://www.levinecentral.com/ham/grid_square.php

SSTV:
https://amsat-uk.org/beginners/iss-sstv/

AMSAT:
https://en.wikipedia.org/wiki/AMSAT

ARISS:
http://www.ariss.org

OSCARs:
https://en.wikipedia.org/wiki/Amateur_radio_satellite

Emerald Time iPhone Time App:
https://itunes.apple.com/us/app/emerald-
time/id290384375?mt=8

Arrow II 146/437-10WBP Antenna

http://www.arrowantennas.com/arrowii/146-437.html

Elk 2M/440L5 Antenna
https://elkantennas.com/product/dual-band-2m440l5-log-periodic-antenna/

$4.00 Ham Radio Satellite Antenna (Youtube Video)
https://www.youtube.com/watch?v=Hy_XwvMmIro

ISS Fan Club
http://issfanclub.com/

ISSTracker (no predictions, just live tracking)
http://www.isstracker.com/

A very good web based site for satellite prediction is:
http://www.heavens-above.com/

ABOUT THE AUTHOR

Brian Schell (KD8OTD) is a former College IT Instructor who has an extensive background in computers dating back to the 1980s. Currently, he writes on a wide array of topics from computers, to world religions, to ham radio, and even releases the occasional short horror tale.

He'd love to hear your stories of success and failure with FM satellites and antennas. If there's something you would like to see in a future edition of the book, or otherwise have suggestions, please drop him a note. Contact him at:

Web: http://BrianSchell.com
Email: brian@brianschell.com

twitter.com/BrianSchell

facebook.com/Brian.Schell

instagram.com/brian_schell

pinterest.com/brianschell

STAY UP TO DATE!

Join my email update list— There's NO weekly SPAM or filler material, only announcements of new books or major updates.

http://brianschell.com/list/

HELP ME!

Contact the Author

If you have a suggestion or find a mistake, email me about it, and I'll get it into the next edition of the book. Got a gripe, complaint, question, or just adoring fan mail? Same thing!

Leave a Review

If this book helped you, please leave a review where you purchased this book. Reviews are the best way to help out!

Share With Your Friends

Did you enjoy this book? Please use the buttons below to spread the word to your friends and followers.

ALSO BY BRIAN SCHELL

Made in the USA
Monee, IL
16 November 2019

16928597R00039